Butterflies

Evoneo Berti Filho
João Angelo Cerignoni

Evoneo Berti Filho
João Angelo Cerignoni

Departament of Entomology
and Acarology – Esalq/USP

Butterflies

Fealq

Av. Centenário, 1080
13416-000 Piracicaba, SP, Brasil

Phone: 19-3417-6600

livros@fealq.org.br
fealq.org.br

Catalogação na Publicação
DIVISÃO DE BIBLIOTECA - DIBD/ESALQ/USP

Berti Filho, Evoneo

Butterflies / Evoneo Berti Filho e João Angelo Cerignoni. - - Piracicaba: FEALQ, 2023.
88 p. : il.

Título original: Borboletas
ISBN: 978-65-89722-20-5

1. Borboletas - Brasil I. Cerignoni, J. A. II. Título

CDD 595.78

Elaborada por Maria Angela de Toledo Leme - CRB-8/3359

AUTHORS

Evoneo Berti Filho

Full Professor (Retired) of the Departament of Entomology and Acarology – Luiz de Queiroz College of Agriculture (ESALQ) – University of São Paulo (USP) in Piracicaba, State of São Paulo, Brazil .

Master and Doctor in Forest Entomology – ESALQ/USP.

Post doctor – Biological Control and Forest Entomology – University of California at Berkeley, USA.

João Angelo Cerignoni

Laboratory Technician (Department of Entomology and Acarology of ESALQ /USP), Piracicaba.

Experience in the areas of Biological Control, General and Forest Entomology working mainly in the biology and breeding of Lepidoptera on alternative food (host) plants in artificial breeding sites (butterfly rearing greenhouse). Knowledge exchange in rearing butterflies in captivity through visits to the main butterfly rearing greenhouses of the Brazilian States of Minas Gerais, Paraná, Santa Catarina and São Paulo.

The butterfly rearing greenhouse
is important when one considers
environment conservation,
because it promotes the rescue, the
repopulation and the reintroduction
of species in their *habitat.*

SUMMARY

Foreword 11

Metamorphosis 12

FAMILY LYCAENIDAE

Subfamily THECLINAE

Pseudolycaena marsyas 16

FAMILY NYMPHALIDAE

Subfamily BIBLIDINAE

Callicore sorana 18

Catonephele numilia penthia 19

Diaethria clymena 20

Dynamine agacles 21

Dynamine postverta 22

Hamadryas amphinome 23

Hamadryas februa 24

Hamadryas feronia 25

Temenis laothoe 26

Subfamily CHARAXINAE

Archaeoprepona demophon 27

Subfamily DANAINAE

Danaus gilippus gilippus 28

Danaus plexippus 29

Hypotyris euclea 30

Mechanitis polymnia 31

Methona themista 32

Placidina euryanassa 33

Thyridia psidii 34

Subfamily HELICONIINAE

Agraulis vanillae 35

Dione juno 36

Dryas iulia 37

Eueides isabella dianasa 38

Heliconius erato phyllis 39

Heliconius ethilia narcaea 40

Subfamily NYMPHALINAE

Colobura dirce 41

Siproeta epaphus 42

Siproeta stelenes 43
Smyrna blomfildia 44

SUBFAMILY SATYRINAE
Caligo illioneus 45

FAMILY PAPILIONIDAE
SUBFAMILY PAPILIONINAE
Battus polydamas 48
Heraclides anchisiades capys 49
Heraclides astyalus 50
Heraclides hectorides 51
Heraclides thoas brasiliensis 52
Parides neophilus 53
Pterourus scamander 54

FAMILY PIERIDAE
SUBFAMILY COLIADINAE
Eurema albula 56
Eurema deva 57
Phoebis argante argante 58
Phoebis philea philea 59
Phoebis sennae 60

SUBFAMILY DISMORPHINAE
Dismorphia amphiona astynome 61

SUBFAMILY PIERINAE
Ascia monustes 62
Melete lycimnia 63

BUTTERFLY HOUSE 64

ORGANIZATION OF THE BUTTERFLY BREEDING GREENHOUSE 68

RULES AND LAWS OF IBAMA TO SET UP A BUTTERFLY REARING GREENHOUSE 74

GUIDELINES OF THE SUSTAINABLE MANAGEMENT PLAN FOR REARING INSECTS OF THE ORDER LEPIDOPTERA 78

MOST FREQUENTLY ASKED QUESTIONS 80

STINGING CATERPILLARS 84

ACKNOWLEDGEMENTS 88

FOREWORD

Insects are the most abundant organisms of the Animal Kingdom, for they encompass about 80% of all living organisms in Earth. The coleopterans (Order Coleoptera) form the Order with the highest number of species, while the lepidopterans (Order Lepidoptera), comprising butterflies and moths, is the second Order in number of species.

Lepidoptera are holometabolous insects, that is, they have complete development, presenting the stages of egg, larva (which in Lepidoptera is termed caterpillar), pupa or chrysalis, and adult. The caterpillars have chewing mouth parts while the adults have sucking mouth parts with a proboscis. Due to the chewing habit of the immature stage (caterpillar) and the sucking habit of the adults, the Lepidoptera are called metagnate insects. Although Lepidoptera species are considered pests in agriculture, forestry, and areas so diverse as the food processing ones, they are the most efficient pollinators after the bees.

Nocturnal lepidopteran, the so-called moths, may present venomous caterpillars and pupa inside cocoons. On the other hand the diurnal and vespertine lepidopterans, the so-called butterflies, never present venomous caterpillars or cocoons.

Lepidoptera occur in almost all of the terrestrial ecosystems, mainly in woods and forests of the tropical and subtropical regions, where the diversity of species is high. However the anthropic action has modified and/or destroyed miles of green areas of wood and forests, thus bringing a number of butterly species to risk of extintion.

METAMORPHOSIS

As already referred, the Lepidoptera are holometabolous insects, that is, they present complete metamorphosis. The female lays the eggs on the leaves, flowers, buds or stems of the host plant and after hatching the caterpillars start the larval period feeding on the host plant and shedding their tegument (skin) at each ecdisis as they grow up to reach the pupa and adult stages. During growth caterpillars molt four to five times. The duration of the egg, larva, pupa and adult stages varies among the different butterfly species. See below the complete metamorphosis of *Danaus plexippus erippus*, known as the monarch butterfly. The host plant is *Asclepias curassavica*, popularly known as "milkweed".

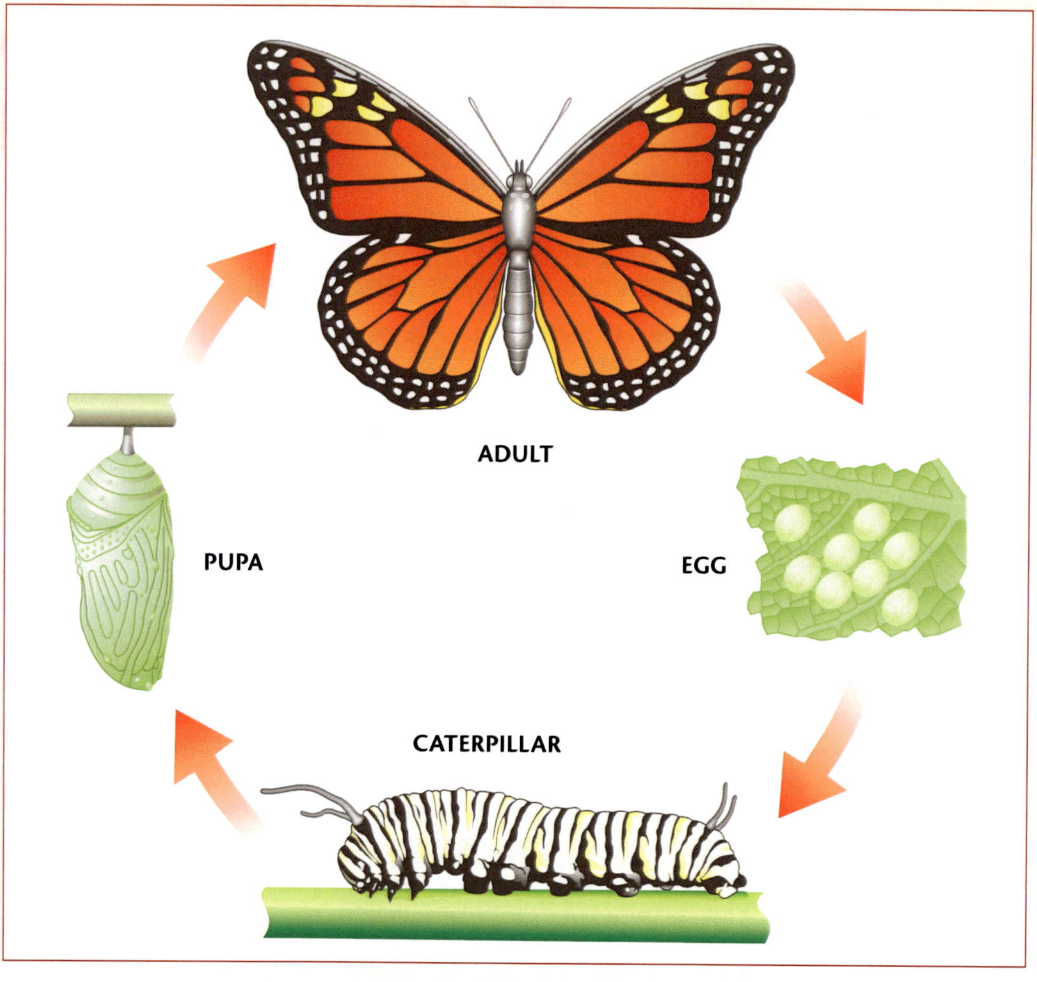

EGG

CATERPILLAR

RECENTLY
FORMED PUPA

MATURE PUPA

ADULT
EMERGENCE

STRECHING THE WINGS

ADULT

FAMILY

LYCAENIDAE

Pseudolycaenas marsyas

PUPA

CATERPILLAR

- **Name:** *Pseudolycaenas marsyas*
- **Family:** Lycaenidae
- **Subfamily:** Theclinae
- **Host plants:** *Erythrina* spp.
- **Biology**

It is estimated that circa of 6,500 Lycaenidae species occur in the world, being 40% in the Neotropical Region, such as the species *Pseudolycaena marsyas* (Lycaenidae, Theclinae). The eggs are laid on the sprouts or young leaves of the host plants *Erythrina* spp., also known as "suinã" or "mochoco" in Brazil. The green caterpillars mimic the host plant leaves. The light brown pupae fix on old leaves. Upon emergence, the adult flies looking for food on fermented fruits or flowers at the wood borders. The wings are metallic blue on the dorsum and ventrally whitish with black points or lines. Characteristically the lycenids present from one to three slender and curved filaments at the end of the rear wings and whose movements simulate the antennas to confuse the predators and thus escaping from them.

FAMILY

NYMPHALIDAE

Callicore sorana

PUPA

VENTRAL DORSAL

CATERPILLAR

- **Name:** *Callicore sorana*
- **Family:** Nymphalidae
- **Subfamily:** Biblidinae
- **Host plant:** *Cardiospermum* sp.
- **Biology**

The caterpillars feed on vines of Euphorbiaceae and Sapindaceae. The ventral face of the rear wings has drawings that resemble the number 80. The host plant occurs in the borders and interior of the woods and the adult female lays the eggs on the tendrils or on sprouts. The caterpillars present bristles on the body, thus confusing the predators. When becomes pupa, it mimics the plant leaves.

Catonephele numilia penthia

NYMPHALIDAE

FEMALE

MALE

PUPA

CATERPILLAR

- **Name:** *Catonephele numilia penthia*
- **Family:** Nymphallidae
- **Subfamily:** Biblidinae
- **Host plant:** *Alchornea cordata*
- **Biology**

This species presents sexual dichromatism and dimorphism. The adult male has six yellow rounded spots on the dorsum of the wings, while the female has white ones on the dorsum of the anterior wings and red ones on the dorsum of the posterior wings. They occur in high numbers in the autumn, looking for *Alchornea* plants growing on the wood border, for the females hardly lay eggs on plants in open fields. The caterpillars present bristles resembling the venomous ones due to their frightful aspect. They eat the tip of the leaf and rest on the central rib where they are hardly noticed. The dark green colored pupa mimics the host plant leaf.

Diaethria clymena

VENTRAL

DORSAL

PUPA

CATERPILLAR

- **Name:** *Diaethria clymena*
- **Family:** Nymphalidae
- **Subfamily:** Biblidinae
- **Host plant:** *Trema micrantha*
- **Biology**

This species is known as the 88 butterfly (notice the resemblance with the number on the ventral surface of the posterior wings). This is a common insect in most environments, including the anthropic ones. It flies in open and sunny places and feeds on falling fruits. People believe that this butterfly brings good luck for those who see it.

Dynamine agacles

PUPA

CATERPILLAR

NYMPHALIDAE

- **Name:** *Dynamine agacles*
- **Family:** Nymphalidae
- **Subfamily:** Biblidinae
- **Host plant:** *Dalechampia scandens*
- **Biology**

This is a species hard to rear un laboratory, but presents a fair development in the field inside caged host plants. The caterpillars remain on the flower sprouts feeding on them. Although innocuous, they have a frightful aspect because they present bristles resembling venomous caterpillars. Notice, however, that when dealing with moths, the bristles may indicate that the caterpillar is venomous.

Dynamine postverta

PUPA

CATERPILLAR

FEMALE

MALE

- **Name:** *Dynamine postverta*
- **Family:** Nymphalidae
- **Subfamily:** Biblidinae
- **Host plant:** *Dalechampia scandens*
- **Biology**

This is also a species hard to rear in laboratory, but presents a fair development in the field inside caged host plants. The caterpillars remain on the flower sprouts feeding on the flowers. Although innocuous, they have a frightful aspect because they present bristles resembling venomous caterpillars. The adults present sexual dichromatism.

Hamadryas amphinome

NYMPHALIDAE

PUPA

CATERPILLAR

- **Name:** *Hamadryas amphinome*
- **Family:** Nymphalidae
- **Subfamily:** Biblidinae
- **Host plant:** *Delachampia scandens*
- **Biology**

This species is called "cracker butterfly" due to the sound it produces when flying and it is a species easy to rear in laboratory species. It is interesting to observe the way adults rest with the head downwards and the wings extended mimetizing the tree trunk bark and lichens. They measure circa 7.5 cm of wingspan. The pupa is similar to a twisted dry leaf with two projections at the apex.

Hamadryas februa

PUPA

CATERPILLAR

- **NAME:** *Hamadryas februa*
- **FAMILY:** Nymphalidae
- **SUBFAMILY:** Biblinidae
- **HOST PLANT:** *Delachampia scandens*
- **BIOLOGY**

As *H.amphinome*, it is also a species easy to rear in laboratory and their caterpillars have a frightful aspect because they present bristles resembling venomous caterpillars.

Hamadryas feronia

NYMPHALIDAE

PUPA

CATERPILLAR

- **Name:** *Hamadryas feronia*
- **Family:** Nymphalidae
- **Subfamily:** Biblidinae
- **Host plant:** *Delachampia scandens*
- **Biology**

As *H.amphinome* and *H.februa*, this is also a species easy to rear in laboratory. However, there is a problem with the pupa which may be attacked by a hymenopteran parasitoid. Concerning the three species of *Hamadryas* the immature stages may be gregarious as in *H.amphinome*, or solitaries as in *H.februa* and *H.feronia*.

Temenis laothoe

PUPA

CATERPILLAR

- **Name:** *Temenis laothoe*
- **Family:** Nymphalidae
- **Subfamily:** Biblidinae
- **Host plant:** *Cardiospermum* sp.
- **Biology**

As stated to the species *Callicore sorana*, the caterpillars of *T.laothoe* also feed on vines of Euphorbiaceae and Sapindaceae. The host plant occurs in the borders and interior of the woods and the adult female lays the eggs on the tendrils or on sprouts. The caterpillars present bristles on the body, thus confusing the predators. When it becomes a pupa it mimics the plant leaves.

Archaeoprepona demophon

PUPA

CATERPILLAR

NYMPHALIDAE

- **Name:** *Archaeoprepona demophon*
- **Family:** Nymphalidae
- **Subfamily:** Charaxinae
- **Host plant:** *Annona squamosa*
- **Biology**

Adults present rapid, irregular and always high fly briefly resting to feed on mature or fermented fruits. The eggs are laid on plants of the Annonaceae family, placing one single egg on tender tendrils or sprouts. Because it is a fairly unusual insect, one hardly find a number of caterpillars on the same host plant. The caterpillar may measure up to 8 centimeters and uses to feed at night, for during the day it mimics the dry foliage. It is light brown on the dorsum and a lighter brown on the thorax, with appendages on the head resembling horns which are used as defense when the species is disturbed. The biological cycle is long and the pupa is green like the green leaves of the host plant. The adult is dark with a blue strip on the dorsal center of the wings while the ventral face is clair. It is hard to rear this genus in a butterfly rearing greenhouse, for the adults are pretty kinky and difficultly adapted to a butterfly rearing greenhouse.

Danaus gilippus gilippus

PUPA

CATERPILLAR

- **Name:** *Danaus gilippus gilippus*
- **Family:** Nymphalidae
- **Subfamily:** Danainae
- **Host plant:** *Asclepias curassavica*
- **Biology**

This species may be found in most different wild places. The light green eggs are isolately laid on the pointers of the host plant. The caterpillar, brown and yellow colored with white strips in between, measures up to 7 centimeters and the head is black with small clear rings. The barrel shaped pupa is light green with golden dots in the central area. The adults visit the flowers of *Lantana camara* and *Asclepias curassavica*. However, one should be aware that the caterpillars host plants are *Asclepias curassavica* and *Sarcostemma* sp. (Asclepiadaceae).

Danaus plexippus

PUPA

CATERPILLAR

NYMPHALIDAE

- **Name:** *Danaus erippus*
- **Family:** Nymphalidae
- **Subfamily:** Danainae
- **Host plant:** *Asclepias curassavica*
- **Biology**

It measures circa 9.5 centimeters of wingspan. Its flying is low and slow in open places and the species presents migratory habits. The eggs are laid on flower buds or under the leaves of the toxic plant *Asclepias curassavica* popularly known as "milk weed" ("official-de-sala", in Brazilian Portuguese). Regarding egg hatching, the caterpillars present yellow and black strips and feed on the leaves of the toxic host plant thus becoming unpalatable to predators (like the birds for instance), but still being parasitized by certain wasps and mainly by Tachinidae fly parasitoid species. The rearing requires care because the caterpillars are extremely susceptible to virus diseases. One should also be careful when collecting large caterpillars, for most of them may be parasitized by the already mentioned Tachinidae flies. This is a nice butterfly that should be considered as a one of the species selected to compose a rearing butterfly house.

Hypothyris euclea

PUPA

CATERPILLAR

- **Name:** *Hypothyris euclea*
- **Family:** Nymphalidae
- **Subfamily:** Ithomiinae
- **Host plant:** *Acnistus arborescens*
- **Biology**

Adults fly in the under bushes and wood borders, mating and looking for the host plant *Acnistus arborescens* ("fumo-bravo" in Brazilian Portuguese) on which they lay their eggs. The caterpillars present black strips on the white body and are extremely voracious with a rapid larval stage. They usually pupate in the dry branches of the host plant.

Mechanitis polymnia

PUPA

CATERPILLAR

NYMPHALIDAE

- **Name:** *Mechanitis polymnia*
- **Family:** Nymphalidae
- **Subfamily:** Ithomiinae
- **Host plant:** *Solanum paniculatum*
- **Biology**

This is a very common Brazilian species measuring circa 6.5 centimeters of wingspan. It presents a slow flight normally near the soil surface. The caterpillar is white with red head. The pupa is silvery and reminds a mirror. It is an excellent species to be selected for an intensive rearing. It is easy to manipulate for practically there are no problems with diseases, nor predators and parasitoids.

Methona themisto

PUPA

CATERPILLAR

- **Name:** *Methona themisto*
- **Family:** Nymphalidae
- **Subfamily:** Ithomiinae
- **Host plant:** *Brunfelsia hopeana*
- **Biology**

This species is popularly known as caterpillar of manacá ("lagarta-do-manacá", in Brazilian Portuguese). Adults present a light and constant flight visiting the host plant garden manacá ("manacá-de-jardim", in Brazilian Portuguese) where the females lay their eggs isolated on sprouts or young leaves. The caterpillars are black with yellow rings between the body segments. They leave the host plant to pupate in the surroundings.

Placidina euryanassa

PUPA

CATERPILLAR

NYMPHALIDAE

- **Name:** *Placidina euryanassa*
- **Family:** Nymphalidae
- **Subfamily:** Ithomiinae
- **Host plant:** *Brugmansia suaveolens*
- **Biology**

Adults present a slow and low flight in the wood borders (forest edges), looking for the host plant *Brugmansia suaveolens* ("trombeteira" in Brazilian Portuguese), on which the females lay the egg masses. The caterpillars, black with a red head, usually remain together during the day in hiding places near the host plant. During the night they leave these places to feed on the plants. When the sun rises they return to their hiding places. To pupate they leave the plant fixing on the dead vegetation.

Thyridia psidii

PUPA

CATERPILLAR

- **Name:** *Thyridia psidii*
- **Family:** Nymphalidae
- **Subfamily:** Ithomiinae
- **Host plant:** *Cyphomandra betaceae*
- **Biology**

The adult resembles the ones of *Methona themisto.* Their flight is light and constant visiting Solanaceae and laying their eggs, isolately, on sprouts or young leaves of arboreal tomato plants. The caterpillars are white such as those of *Mechanitis lysimnia* although larger measuring 6 centimeters. The pupa, as well as that of *Mechanitis*, is silvery and larger than the *Mechanitis* one.

Agraulis vanillae

PUPA

CATERPILLAR

NYMPHALIDAE

- **Name:** *Agraulis vanillae*
- **Family:** Nymphalidae
- **Subfamily:** Heliconiinae
- **Host plant:** *Passiflora suberosa*
- **Biology**

This is a common species measuring circa of 7 centimeter of wingspan. Its flight is rapid, low and irregular. The eggs are laid isolately on young leaves. It is considered an insect pest of passion fruit. The caterpillar is brown with black spines scattered along the body. The species is known as silver drop butterfly due to the silvery spots on the ventral face of the wings.

Dione juno

PUPA

CATERPILLAR

- **Name:** *Dione juno*
- **Family:** Nymphalidae
- **Subfamily:** Heliconiinae
- **Host plant:** *Passiflora* spp.
- **Biology**

This butterfly measures 6 centimeters of wingspan. Its flight is rapid and hard to follow and at nightfall they gather in small groups on the same plant. The eggs are laid on the shoots or on the bifurcation of leaves of several species of passion fruit. The caterpillar, measuring up to 4 centimeters, is brown, prickly and thorny and feeds on the leaves of passionfruit (*Passiflora* sp.). Cannibalism may occur when food is scarce. It is not easy to maintain a laboratory rearing for diseases may cause total loss of the insect colony. However, this problem may be prevented by maintaining the caterpillars on caged host plants in the field, thus also protecting them against the action of predators and parasitoids.

Dryas iulia

PUPA

CATERPILLAR

NYMPHALIDAE

- **Name:** *Dryas iulia*
- **Family:** Nymphalidae
- **Subfamily:** Heliconiinae
- **Host plant:** *Passiflora mucronata*
- **Biology**

This species measures circa 8 centimeters of wingspan and is very common visiting flowering plants and gardens. Its flight is rapid, low and irregular. The eggs are isolately laid on dry tendrils. The caterpillar is brown with long black bristles scattered along the body. This butterfly is considered one of the insect pests of passionfruit.

Eueides isabella dianasa

PUPA

CATERPILLAR

- **Name:** *Eueides isabella dianasa*
- **Family:** Nymphalidae
- **Subfamily:** Heliconiinae
- **Host plant:** *Passiflora edulis*
- **Biology**

This is a very common species found in passion fruit either in plantings or in the wilds. The caterpillar is dark, with yellow strips along the body. In field conditions it is common to be preyed in the larva stage and parasitized in the pupa stage. However it is quite easy to be reared in laboratory.

Heliconius erato phyllis

PUPA

CATERPILLAR

NYMPHALIDAE

- **Name:** *Heliconius erato phyllis*
- **Family:** Nymphalidae
- **Subfamily:** Heliconiinae
- **Host plant:** *Passiflora* sp.
- **Biology**

This butterfly measures circa of 7 centimeters of wingspan. Its flight is slow, and 2 meters high from the soil. At nightfall, they gather into small groups on twigs of the same plant in the rearing greenhouse. The adult is dark, with a large red band on the dorsum of the anterior wing and a yellow stripe on the dorsum of the posterior wing. The eggs are laid on the buds or on the tendrils of several species of passionfruit plants. The 4-centimeter caterpillar is white with black and red small dots and presents bristles. It feeds on the leaves and cannibalism may be observed when food is scarce.

Heliconius ethilla narcaea

PUPA

CATERPILLAR

- **Name:** *Heliconius ethilla narcaea*
- **Family:** Nymphalidae
- **Subfamily:** Heliconiinae
- **Host plant:** *Passiflora* sp.
- **Biology**

This species measures up to 7 centimeters of wingspan. It can be found in wood borders and in cultivated and urban areas. Its flight is slow and at variable heights and they are attracted by flowers. The eggs are laid on passion fruit buds or tendrils. The caterpillar is white, with black and orange dots and presents bristles.

Colobura dirce

NYMPHALIDAE

PUPA

CATERPILLAR

- **Name:** *Colobura dirce*
- **Family:** Nymphalidae
- **Subfamily:** Nymphalinae
- **Host plant:** *Cecropia* sp.
- **Biology**

This species is popularly known as "zebra butterfly" due to the drawings on the ventral face of both wings. Adults feed on mature fruits and seldom leave the shaded areas of the wood borders and the females lay the eggs on the ventral face of young leaves. Upon hatching, the caterpillars remain together and cut the leaf veins possibly to decrease the sap flow or simply to make themselves a shelter. They eat the leaf always upwards and when they move to another leaf they also cut the veins. The caterpillars grow together but separate when leaving the host plant at night to pupate. It is not easy to find their pupae.

Siproeta trayja

PUPA

CATERPILLAR

- **Name:** *Siproeta trayja*
- **Family:** Nymphalidae
- **Subfamily:** Nymphalinae
- **Host plant:** *Ruelia* sp.
- **Biology**

The caterpillars are black with black bristles laterally and yellow ones on the dorsum. As they feed at night, it is hard to find them during the day on the host plant. The pupa is greenish, turning dark upon adult emergence. Adult is dark with a white strip on the wings.

Siproeta stelenes

PUPA

CATERPILLAR

NYMPHALIDAE

- **Name:** *Siproeta stelenes*
- **Family:** Nymphalidae
- **Subfamily:** Nymphalinae
- **Host plant:** *Ruelia* sp.
- **Biology**

The adult, black with green large areas on the wings, is one of the most spectacular specimens of the Nymphalinae, flying in the sunny open areas of the woods, feeding on fermented fruits and looking for partners to mate. The caterpillars are black with black bristles laterally and orange ones on the dorsum; they feed at night and therefore are hardly found on the host plant during the day.

Smyrna blomfildia

VENTRAL

DORSAL

PUPA

CATERPILLAR

- **Name:** *Smyrna blomfildia*
- **Family:** Nymphalidae
- **Subfamily:** Nymphalinae
- **Host plant:** *Urera baccifera*
- **Biology**

The caterpillars feed on *Urera baccifera* (Urticaceae) and generally occur in the summertime. The females spread their eggs on young leaves or sprouts. The caterpillars are dark on hatching, then turn green and finally brown when near to pupation. Because caterpillars present pine-like spines they seem to be venomous, but they are innocuous. Adults emerge in about eight days and are brown colored with a yellow strip on the middle of the anterior wing with three white spots on the black apex of this wing.

Caligo illioneus

VENTRAL

DORSAL

PUPA

CATERPILLAR

NYMPHALIDAE

- **Name:** *Caligo illioneus*
- **Family:** Nymphalidae
- **Subfamily:** Satyrinae
- **Host plant:** *Musa* spp.
- **Biology**

This is one of the largest Brazilian species of butterfly, reaching circa of 18 centimeters of wingspan, and is popularly known as "owl butterfly", because it presents a round spot on the ventral face of each posterior wing resembling owl eyes, with black pupil and clear iris, frightening the predators. The adult lives circa 45 days in the greenhouse. The caterpillar eats the leaf of banana tree and other plants of the family Musaceae and is light brown with two black strips along the body; the head capsule is dark brown.

FAMILY

PAPILIONIDAE

Battus polydamas

PUPA

CATERPILLAR

- **Name:** *Battus polydamas*
- **Family:** Papilionidae
- **Subfamily:** Papilioninae
- **Host plant:** *Aristolochia galeata*
- **Biology**

This insect is found in the wood borders and also in urban environments. Its population is dependent on the availability of its host plants of the genus *Aristolochia*, which generally occurs at the wood borders. However, such plants are destroyed by the anthropic actions for clearing the wood borders in order to increase the cultivation area. The couples fly frenzily in areas with plants presenting a high potential of nectar where they feed and afterwards copulate. The eggs are yellow and laid in groups of three to five on plant sprouts. Upon emergence, the caterpillars usually feed in groups. They are dark colored with spine-like appendages, and as all the Papilionidae, when disturbed they evert the osmeterium, a protractile organ, orange colored on the dorsum of the protoraxic segment, which ejects a fetid odor that repels their natural enemies. They pupate in the same plant and in a few days emerge the adults which are dark colored with yellow lemon spots on the wings and small red spots on the abdomen. Most of the pupae collected in the field are parasitized by hymenopterans.

Heraclides anchisiades capys

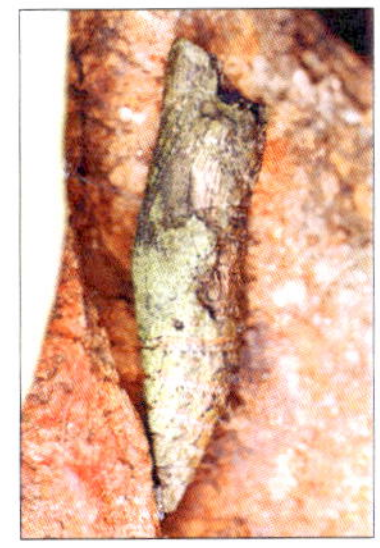

PUPA

CATERPILLAR

PAPILIONIDAE

- **Name:** *Heraclides anchisiades capys*
- **Family:** Papilionidae
- **Subfamily:** Papilioninae
- **Host plant:** *Citrus* spp.
- **Biology**

This species is a little small when compared to most of the Papilionidae. The adults also present sexual dimorphism and dichromatism; the females are dark, with white spots on both anterior and posterior wings and red spots bordering the posterior ones. Males are almost dark colored, presenting yellow strips on both wings and red spots on the posterior ones. Such as *Heraclides astyalus*, the caterpillars do not feed in group and present a relatively slow growth. They are blackish with yellow, white and orange spots mimetizing bird excrements. After pupation the colors turn to brownish and greenish thus mimetizing the plant branching. It was observed that pupae of the same age may undergo a kind of dormancy and the adults emerge some months after the majority of the group. It is probably a survival strategy.

Heraclides astyalus

MALE

FEMALE

PUPA

CATERPILLAR

- **Name:** *Heraclides astyalus*
- **Family:** Papilionidae
- **Subfamily:** Papilioninae
- **Host plant:** *Citrus* spp., *Esenbeckia febrifuga*
- **Biology**

This adult species presents sexual dimorphism and dichromatism, being the male a little smaller, yellowish with black strips, resembling *Heraclides thoas brasiliensis*, while the females are blackish with yellow details on the wing borders. When in captivity the females lay the eggs on the *Citrus* spp. sprouts and young leaves. The caterpillars do not feed in group and present a relatively slow growth; they are blackish with yellow, white and orange spots mimetizing bird excrements. After pupation the colors turn to brownish and greenish thus mimetizing the plant branching.

Heraclides hectorides

MALE

FEMALE

PUPA

CATERPILLAR

- **Name:** *Heraclides hectorides*
- **Family:** Papilionidae
- **Subfamily:** Papilioninae
- **Planta hospedeira:** *Citrus* spp.
- **Biology**

This species is a little small when compared to most of the Papilionidae. The adults also present sexual dimorphism and dichromatism; the females are dark with white spots on both anterior and posterior wings and red spots bordering the posterior ones. Males are almost dark colored, presenting yellow strips on both wings and red spots on the posterior ones. Such as *Heraclides astyalus*, the caterpillars do not feed in group and present a relatively slow growth. They are blackish with yellow, white and orange spots mimetizing bird excrements. After pupation the colors turn to brownish and greenish thus mimetizing the plant branching. It was observed that pupae of the same age may undergo a kind of dormancy and the adults emerge some months after the majority of the group. It is probably a survival strategy.

Heraclides thoas brasiliensis

PUPA

CATERPILLAR

- **Name:** *Heraclides thoas brasiliensis*
- **Family:** Papilionidae
- **Subfamily:** Papilioninae
- **Host plants:** *Citrus* spp.
- **Biology**

This is a quite common species in Brazil, measuring 13 centimeters of wingspan. It lives in the woods, in sunny areas and mainly on wood borders during the whole year and when at rest the wings are wide open on the plant. The eggs are laid isolatedly on the sprouts and young leaves of citric plants such as lemon, orange and tangerine. The caterpillars eat the leaves and when threatened, they release an odoriferous substance from the osmeterium. The pupae seem plant branches, fooling the predators. The adult is dark and yellow colored and has a red spot in each posterior wing.

Parides neophilus

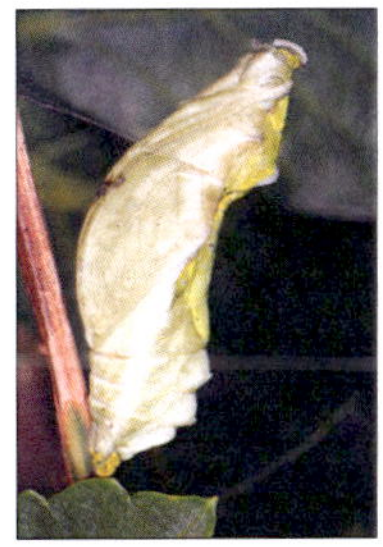

PUPA

CATERPILLAR

PAPILIONIDAE

- **Name:** *Parides neophilus*
- **Family:** Papilionidae
- **Subfamily:** Papilioninae
- **Host plants:** *Aristolochia* spp.
- **Biology**

The adult presents a fast, irregular and always high flight and feeds in flower nectaries at each brief stopping point. The eggs are laid on plants of the genus *Aristolochia* being one single egg on each tender sprout. It is a quite rare insect and one hardly finds several caterpillars on the same host plant. Being somehow frightening the caterpillars are dark brown with dark and yellow spine like protuberances which serve as a defense when they are disturbed. Its life cycle is long, and when reaching a pupa stage the pupa is green and hardly seen among the host leaves. The adult is dark, almost black colored, with blue spots on the front wings and red ones on the rearing winds. This is a hardly to rear in captivity insect.

Pterourus scamander

PUPA

CATERPILLAR

- **Name:** *Pterourus scamander*
- **Family:** Papilionidae
- **Subfamily:** Papilioninae
- **Host plant:** *Persea gratissima*
- **Biology**

This insect has a wingspan of nearly 12 centimeters. Its flight is rapid and regular at a mean height of 4 meters. The eggs are scarcely laid on avocado young leaves. The caterpillar is green with a dark brown X on the dorsum. This is also a species hard to rear in laboratory.

FAMILY

PIERIDAE

Eurema albula

PUPA

CATERPILLAR

- **Name:** *Eurema albula*
- **Family:** Pieridae
- **Subfamily:** Coliadinae
- **Host plant:** *Senna occidentalis*
- **Biology**

The adult is small, measuring 3.5 cm of wingspan. They are white with a black blotch at the tips of the front wings. They are found in the wild under bushes and crop areas flying slowly at circa one meter from the ground level. The eggs are laid on the sprouts or on young leaves of *Senna occidentalis* or *Senna hirsuta* where the green caterpillars are confused with the greenish of the plants. This is a species easy to rear in butterfly houses.

Eurema deva

PUPA

CATERPILLAR

PIERIDAE

- **Name:** *Eurema deva*
- **Family:** Pieridae
- **Subfamily:** Coliadinae
- **Host plant:** *Senna occidentalis*
- **Biology**

The yellowish colored adults measure 5 cm of wingspan, fly rapidly from the ground level up to 5 meters high. The eggs are laid on sprouts or on young leaves of *Senna occidentalis* and the green caterpillars are hardly noticed on the greenish parts of the host plant. This species, like other Pieridae, fly in high numbers on puddles or humid ground where substances rich in nitrates help the sexual maturation of the males.

Phoebis argante argante

PUPA

CATERPILLAR

- **Name:** *Phoebis argante argante*
- **Family:** Pieridae
- **Subfamily:** Coliadinae
- **Host plant:** *Inga uruguensis*
- **Biology**

A very common species found in woods, crops, and urban areas. It presents a rapid fly while looking for flowers. The cylindrical and sharp eggs are laid on the young buds. Other species of its host plant also present floral nectarines (nectar producing gland) which feeds some ant species. These ants protect the plants against enemies and the eggs, considered such as them, are destroyed as soon as they are laid. Therefore if one wants to rear this species he would have to protect its eggs against the ants. The green caterpillars and pupae are mimetic with their habitat.

Phoebis philea

PIERIDAE

PUPA

CATERPILLAR

- **Name:** *Phoebis philea*
- **Family:** Coliadinae
- **Subfamily:** Pierinae
- **Host plant:** *Cassia siamea*
- **Biology**

The wingspan reaches up to 9 cm. Normally is seeing flying from the ground level to the highest crown trees. The eggs are laid on the sprouts or young leaves of the host plant. It is quite difficult to catch the adult due to its very rapid fly. The male is yellow as the egg yolk and the female is clearer yellow.

Phoebis sennae

PUPA

CATERPILLAR

- **NAME:** *Phoebis sennae*
- **FAMILY:** Pieridae
- **SUBFAMILY:** Coliadinae
- **HOST PLANT:** *Cassia* sp.
- **BIOLOGY**

The wingspan reaches up to 10 cm. It usually has a rapid flying in gardens and woods, from the ground level to the highest crown trees. The eggs are laid on sprouts or young leaves of *Cassia* sp. When at rest, the caterpillars are hardly seen because they present the same colors of the host plant. The pupa is characteristic of the Pieridae and the adults fly together. Such behavior occurs in the beginning of the male adulthood, when the great groups feed on pools or humid grounds rich in nitrates which help the sexual maturation of the males.

Dismorphia amphiona astynome

PUPA

PIERIDAE

CATERPILLAR

- **Name:** *Dismorphia amphiona astynome*
- **Family:** Pieridae
- **Subfamily:** Dismorphiinae
- **Host plant:** *Inga uruguensis*
- **Biology**

This species was never observed being reared in butterfly houses. It presents an outstanding mimetism with *Mechanitis* sp. (Nymphalidae, Ithomiinae) and probably females collected at the end of the autumn were considered as *Mechanitis* sp. and released in the butterfly house where they laid their eggs on the tips and young leaves of *Inga*. Thereafter the adults were identified as *Dismorphia amphiona astynome.*

Ascia monuste

PUPA

CATERPILLAR

- **Name:** *Ascia monuste*
- **Family:** Pieridae
- **Subfamily:** Pierinae
- **Host plant:** *Brassica oleracea*
- **Biology**

This is a very common species in Brazil, measuring 6 cm of wingspan and being considered as one of the key-pests of collard green. The yellow colored eggs are laid on the leaves of collard green as well as on other Cruciferae. It is easily reared in butterfly houses.

Melete lysimnia

PUPA

CATERPILLAR

PIERIDAE

- **Name:** *Melete lysimnia*
- **Family:** Pieridae
- **Subfamily:** Pierinae
- **Host plant:** *Phoradendron affine*
- **Biology**

This insect usually occurs most frequently in the spring, in the summer and in the beginning of the autumn. The eggs collected in the field present 80% of viability, for part of them may be attacked by the fungus *Metarhizium anisopliae*. From pupa collecting in the field, one may observe the occurrence of the parasitoid *Brachymeria* sp. (Hymenoptera, Chalcididae).

BUTTERFLY HOUSE

Butterfly house refers to a space of confinement similar to a greenhouse aiming at the rearing of butterflies to study their biology, the maintenance of the insect populations and also the development of rearing techniques for obtaining the highest possible number of adults free from the action of natural enemies. Such space (that meets the best conditions for butterfly rearing) should be laterally covered with plastic dark screen (sombrite) on the top to allow sunlight exposition inside the greenhouse, thus providing a natural environment to maintain host plants for the caterpillars and nectar producing ones for the adult butterflies. For this rearing to become self-sustaining it is necessary that a team follows the biological cycle of each species from the eggs, collected on the host plant and taken to the nursery, up to the adult emergence. The adults will be released in the butterfly house to restart the cycle. It is important to emphasize that the team must keep the good conditions inside the butterfly house, by preserving the plants health and avoiding the occurrence of natural enemies. In another space also screen covered one should maintain potted host plants to be moved to the butterfly house whenever needed for the females lay their eggs. When butterfly populations are kept in a greenhouse one observes free individuals flying around it and there-fore it is important to stress that host and attractive plants must surround the butterfly house, for this is an opportunity to collect rare and/or hardly seen species to start a caged rearing. This is a nice way to rescue such species and determine their host plants or even alternative ones, that are plants of the same genus or family well accepted by the caterpillars and not interfering in the insect biological cycle. The family Papilionidae is the most studied one so far. The species of the genera *Heraclides, Pterourus, Battus,* and *Parides* may be used for restocking, and for this purpose groups of their host plants should be kept in the woods because man increasing invasion firstly destroy the habitat of these plants. On the other hand, plants of the genus *Aristolochia* allows the spreading not only of these spe-cies but also of those ones of the genera *Battus* and *Parides*. This could also be done with host plants of the genus *Esenbeckia* and other ones of the family Rutaceae regarding the butterflies of the genera *Heraclides* and *Pterourus*.

Easy rearing species

Easy rearing species are those that when reared inside greenhouses showed rusticity in the immature stage and high longevity in the adult stage. One should study the biology of the highest possible number of species and list those easy to rear in captivity. Even in plastic screen protected in the butterfly house, it is possible to observe natural enemies such as predators and parasitoids, attacking the different life stages of the butterflies, as wasps parasitizing eggs and pupae, heteroptera sucking caterpillars and pupae, even lizards, spiders and birds preying all the life stages (egg, caterpillar, pupa, adult).

The structure

The area where the butterfly house will be built should be chosen according to the north south disposition of the grounds. At east there must have native or planted woods. The structure must be covered by plastic dark screen. Two antechambers, an entrance and an exit ones, also covered with dark screen (sombrite), must be done in order to prevent the input of natural enemies. Plants with high nectar content for adult feeding and potted trap host plants should be kept inside the butterfly house. The females will lay their eggs on these trapped host plants. The eggs are collected and taken to the nursery where the species will be reared up to the adult stage and then being ready to be released in the butterfly house.

Self-sustaining rearing

In order to reach a self-sustaining rearing, a team must keep up with the whole biological cycle of each species from egg collecting up to the release of the adults in the butterfly greenhouse. By doing this, one will have a high number of butterflies which will be separated in couples to be released in the greenhouse while the excess will be released in the free surrounding environment to contribute for the butterfly fauna in the area and/or region.

The team

The properly trained team is responsible for taking the butterfly greenhouse by maintaining the plants from the internal and external areas of the greenhouse, providing fermented fruits for adult feeding, eggs collecting, dead, adults removal, and mainly monitoring the natural enemies. Team work is developed following a rearing flowchart and this works for all the species chosen according with the time of occurrence. Open field collections are made not only to begin a rearing but also to genetically improve captive populations by introducing wild material.

Natural occurrence

After the implantation of the butterfly greenhouse and the butterfly garden (that is, the surrounding garden with host plants of many species as well as plants to provide nectar to the adult species) one will observe lots of flying butterflies of different species, specially the rare ones, visiting these plants. This is a consequence of the phero-mones released by the butterflies flying inside the greenhouse; in addition the variety of flowers, some of them specific for many spe-cies, will certainly make the area very attractive to these butterflies. The occurrence of these rare species will gradually increase because the area provides food for their caterpillars. However one should be aware of the increasing number of natural enemies following the growth of such rare species.

Rare species

To rescue species considered rare in the region, probably due to the deforestation of their area of reproduction, it is necessary to pay attention to the host plants of these species. Such plants should be cultivated inside and outside the greenhouse, mainly at the wood edges where they would be handled to attract these species of rare butterflies. Besides the host plants, alternative ones should be considered, for many species of butterflies may feed on more than one plant species. These alternative plants should not be of the same genus or same family but even so they serve as food for the caterpillars, like those of some Papilionidae species.

Rare Papilionidae

Considering the lack of host plants for some butterfly species either in agricultural and/or urban areas, a number of species will be seen in places such as those where the native vegetation is quite abundant and in clearings in the woods. Among them there can be mentioned those of the family Papilionidae, including the genera *Battus, Heraclides, Parides,* and *Pterourus.* The number of these species will gradually increase as they find the plants where they lay their eggs in the garden surrounding the butterfly greenhouse.

Plant collection

Papilionidae of the genera *Battus* and *Parides* usually lay their egg on plants of the genus *Aristolochia.* The butterfly greenhouse of the Luiz de Queiroz College of Agriculture (Escola Superior de Agricultura Luiz de Queiroz), University of São Paulo, in Piracicaba, State of São Paulo, Brazil, maintain a collection of the species *Aristolochia gigantea, A.fetida, A.macroura, A.elegans,* and also *Passiflora* spp., the host plants of Nymphalidae species. There are also collections of plants selected by scientific researches besides material collected in field trips. Every time one discovers caterpillars on plants from other areas he must collect such plants to be cultivated inside the greenhouse to feed these caterpillars.

ORGANIZATION OF THE BUTTERFLY BREEDING GREENHOUSE

The butterfly breeding greenhouse is quite important for the identification of species and the evaluation of the potential ones, their biological aspects for the breeding and maintenance of the population. Such greenhouses also provide basic data for researches on the Brazilian butterflies for little is known concerning the knowledge of these native ones. Therefore one has information about potential species to breeding and producing butterflies in the different seasons of the year.

Butterfly breeding greenhouse of the "Fundação Zoobotânica" in Belo Horizonte, State of Minas Gerais, Brazil.

Breeding

The methods used in laboratory to rear insects are adapted for the breeding of butterflies, in order to improve the system of mass insect breeding to be implanted in the butterfly breeding greenhouse. The practical part of mass rearing can be summarized in the following topics:

1) Observation of the laying site and the host plant.

2) The eggs are collected, taken to the laboratory breeding and placed in a chamber with controlled conditions of temperature, humidity, and luminosity.

3) The caterpillars upon pots hatching are place in plastic containers (pots) or cages.

4) The caterpillars are fed with the host plant that is the same as the adults do the egg laying

This is how the breeding begins, as well as the evaluation of the biological aspects on the studied species. The steps 1 and 2 can be discarded if the caterpillars are already feeding on the host plant.

To evaluate potential species characteristics such as ease of breeding, longevity of adults in the butterfly breeding greenhouse and occurrence of diseases and/or parasitoids are observed. Therefore it is possible to define which are the main ideal species and which have the potential to be breed. Observations are also made about the butterflies that appear flying near the host plants garden. Aspects such as flight type and height can be described after this visualization.

Pots and Cages

Pots refers to transparent plastic recipients which can be large (20 cm diameter; 10 cm high) or small (6.5 cm diameter; 8 cm high) depending on the size of the caterpillar; when bred singly the caterpillar grows faster than those bred together. The number of caterpillars per pot is quite important, because certain species are cannibalistic. This behavior also appears when the plant food is not supplied to the caterpillars. The cages are used at the stage where the caterpillars are larger in number and have already passed through the breeding pots. Usually caterpillars pupate on the top (lid) of the pot or on the leaf used in its feeding

The cages can also be used for caterpillars that do not require much care since the first instars, as in the case of some species, such as:

- *Heraclides thoas brasiliensis,*
- *Battus polydamas,*
- *Ascia monuste,*
- *Placidina euryanassa,*
- *Heraclides anchisiades capys,*
- *Hamadryas amphinome,*
- *Hamadryas feronia,*
- *Mechanitis* sp.,
- *Hypothiris* sp.

The cages can also vary in size, according to the butterfl y species. The most commonly used size is shown in Figure 1. The larger fi xed cages are maintained inside the breeding greenhouse (Figure 2).

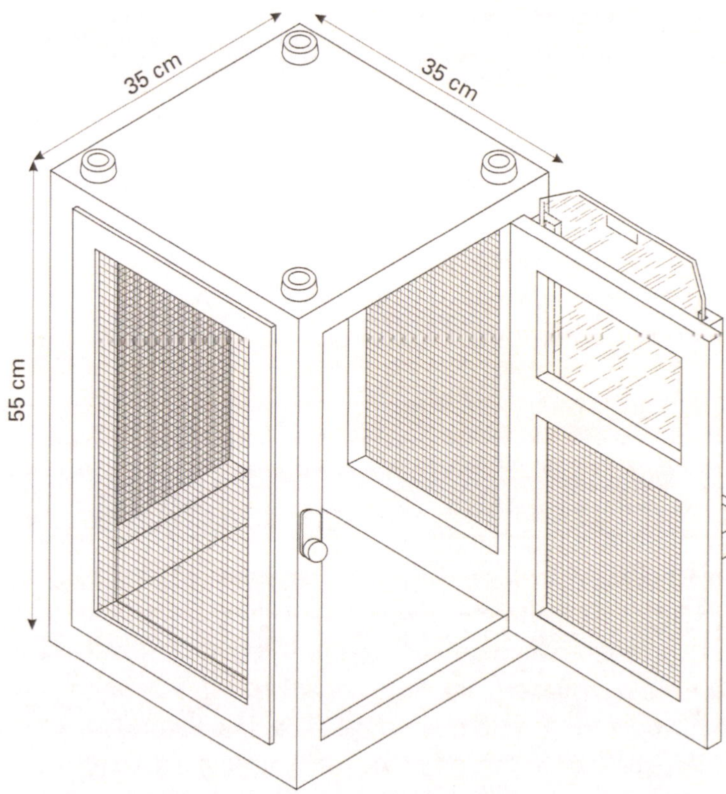

Figure 1. Cage to breeding caterpillars (35 cm wide, 35 cm deep and 55 cm high). The top of the door is glass; the sides have plastic screen; the background and the ceiling have white painted wood

Figure 2. Fixed cage for breeding (model Esalq).

Feeding of caterpillars and butterflies

The caterpillars should be fed with their host plant. Some species have more than one host plant and in this case the caterpillars must be fed with the plant where the laying was done.

The young caterpillars are fed with young host plant leaves because they only scrape the leaf.

From the second or the third instars on the caterpillars are fed with older leaves.

The leaves must be washed in clean water before given to the caterpillars.

The adult butterflies fed on nectar and on fruit juices or pieces place in feeders either metal or plastic ones.

Puparium

The unfixed pupae are coiled by the lower part in cotton and pinned on a Styrofoam plate inside the puparium (27 cm wide, 27 cm high, 26 cm deep). The puparium must be sprayed daily with clean water to maintain the humidity necessary for the emergence of adults.

Diseases and parasitoids

Sick individuals are discarded so they do not infect others in the laboratory. On the other hand the parasitoids can be identified and provide data for biological control studies.

Plants with flowers for the production of nectar inside the butterfly breeding greenhouse.

Metal feeder with butterflies on fruit pieces.

Owl butterflies feeding on fruit pieces.

Parasitoid and the exit hole on a butterfly pupa.

Caterpillar attacked by a parasitoid.

Butterfly rearing greenhouse surrounded by the woods.

RULES AND LAWS OF IBAMA TO SET UP A BUTTERFLY REARING GREENHOUSE

ORDINANCE N. 2314 OF NOVEMBER 26, 1990

The President of the Brazilian Institute of Environment and Renewable Resources (Instituto Brasileiro do Meio Ambiente e dos Recursos Naturais Renováveis – IBAMA) according to the attributions conferred by the law # 7735 of February 22, 1989, published in the Government Official Diary of February 23, 1989 and having regard to the provisions of Article 6th of the Law # 6197, of January 3, 1987, decides:

Article 1st – To institute nurseries to reproduce insects of the Order Lepidoptera from the wild fauna with economic purposes. Unique paragraph – The registration of the breeding grounds, object of this article, will be done by the Control and Inspection Board in accordance with the pertinent legislation after approval of the technical project by the Ecosystem Supervisor (Board of Supervisors).

Article 2nd – For the purpose of this ordinance, one considered breeding sites specially designated areas with facilities capable of enabling the life and breeding of wildlife species where they can receive adequate assistance. Unique paragraph – The breeding grounds must follow the instructions established by the attached Guidelines for the Management Plan.

Article 3rd –The breeding grounds will be classified in the categories of farms managed by farmers and by companies with the objective of managing insects of the Order Lepidoptera in the metamorphisis phase (caterpillar), aiming at the production of butterflies according to the attached guidelines for the Management Plan. Paragraph 1st Breeding grounds managed by rural producers are considered as such when administered by private persons. Paragraph 2nd Breeding grounds managed by companies are considered as such when administered by duly constituted legal persons.

Article 4th – People interested in obtaining registration of a breeding ground with economic purpose at IBAMA ((Instituto Brasileiro do Meio Ambiente e dos Recursos Naturais Renováveis) in the form of this Ordinance shall present:

a) qualification of the private or legal person;

b) location of the enterprise with characterization of the areas, domain, forms of access and general description;

c) schematic sketch of the ground breeding accompanied by delimitation of the domain boundaries, dimensions of the area of attraction/reproduction of insects of the Order Lepidoptera;

d) schematic sketch of the nursery;

e) species of the Order Lepidoptera (diurnal and nocturnal ones) selected to be managed in the property;

f) to relate possible areas (cultivated lands) for collecting Lepidoptera in the caterpillar stage which are outside the area that is object of the "item b" of this article;

g) operational planning of management;

h) technically responsible person, duly qualified with term or contract of work of monitoring and responsibility for the information and technical guidance of the enterprise;

i) declaration that they are aware of the relevant legislation - law # 5.197, of January 3, 1967, law # 7.653, of February 12, 1988, and the present Federal Law.

Article 5th – The repopulation program will be carried out according to the birth result refering to the proportion of males versus females, giving to the breeding the characteristic of self-renewing. Unique paragraph – The breeder may not use females as marketable product.

Article 6th – After compliance with the requirements of this Ordinance and after a technical inspection report by the State Superintendencies of IBAMA, the breeder shall be treated in accordance with article 1st of this Ordinance.

Article 7th – The breeding sites must maintain an updated inventory control file per species for their own control and surveys by IBAMA.

Article 8th – The breeders must submit annual reports with the following data:

a) production by species and sex;

b) number of specimens used in the repopulation program by species and sex;

c) annual stock by species; and

d) quantity marketed per species.

Article 9th – The State Superintendencies of IBAMA will maintain a cadastral register of the breeding sites, which will include, in addition to cadastral information, the control of the exits of the stock through the fixed notes issued where the lepidopteran species and/or code number will be recorded according to the 4th Catalog of the Insects that live in Brazilian plants – Ministry of Agriculture, Department of Defense and Agricultural Inspection, Plant Sanitary Defense Service, Central Laboratory of Plant Pathology – Rio de Janeiro.

Article 10th – The State Superintendencies of IBAMA will only issue the license for transportation of products from the breeding grounds of Lepidoptera when one presents the invoice of the sale.

Article 11 – The renew of registrations will be automatic each year, in accordance with the pertinent legislation, carried out by the Control and Inspection Board after the favorable opinion of the State Superintendency and Ecosystems Board.

Article 12 – Individuals or legal persons that negociate with products or byproducts from lepidopteran breeding sites are obliged to register with IBAMA.

Article 13 – The marketing of products from the breeding sites will be regulated through a specific Ordinance to be published in the Official Gazette of the Brazilian Government.

Article 14 – Failure to comply with the provisions set forth in this Ordinance and the legislation in force will imply warning and, if applicable, the cancellation of registration without prejudice to other applicable legal sanctions.

Article 15 – The missing cases will be resolved by the IBAMA Presidency, after hearing the State Board of Directors and Superintendencies.

Article 16 – This Ordinance takes effect on the date of its publication, being revoked the Ordinances that granted records to nurseries and traders of insects of the Order Lepidoptera.

Tânia Maria Tonelli Munhoz
President

Published in the Official Gazette of Brazilian Government number 239, of December 3, 1990. Section I, p.23.116-7

GUIDELINES OF THE SUSTAINABLE MANAGEMENT PLAN FOR REARING INSECTS OF THE ORDER LEPIDOPTERA

A plan of sustainable management for rearing Lepidoptera must clearly demonstrate that the best way to breed them is to leave it to nature. Interference is only admissible in the last stages of metamorphosis, when numerical control and sexing are necessary. This mode of rearing does not cause damages to nature and the reared species do not have threat of extinction.

The thread of extinction of certain species of Lepidoptera has aspects that should be highlighted, such as lack of plant food for some species in certain regions, which may be caused by climatic factors, devastation of plant cover, mutation of certain species or genetic diseases. Another aspect to be considered is that Lepidoptera whose habitat is the agricultural medium are potencially harmfull to crops for food production and are controlled by agrochemicals. Therefore the management plan for Lepidoptera rearing should be based on outside attraction areas only for cultivated areas because it is understood that forests, woods, etc are natural habitats of butterflies where they must be preserved. This plan should have the following methodology:

- The chosen area, delimited and approved should be transformed into an area of attraction for Lepidoptera and therefore one should have planting of high nectar flowers surrounding the host plants for the several species of Lepidoptera of the region.

- The concentration of host plants, either native or cultivated in the area attracts and facilitates the reproduction of the species. Those interested in the rearing of Lepidoptera should have plant nurseries for replenishment of seedlings
- The biological cycle of a butterfly begins in the area of host plants when the females lay their eggs on the vegetation.
- The area must be protected against natural predators such as ants, spiders, lizards, and rats.
- Onto eclosion the caterpillar feeds firstly on the nutritive contents of the egg shell and thereafter on the leaf of the host plant.
- The development of the caterpillars must be monitored at the various stages of ecdysis (shedding skins) and should be evenly distributed on the host plant leaves.
- Before the last ecdysis the caterpillars must be collected manually to be fed into boxes.
- Near the area one should construct a room where the caterpillars will be placed in wooden boxes covered with thin metallic screen.
- Then the caterpillars must be fed with the host plants up to the pupa stage.
- Adult emergence occurs in the mentioned boxes.
- When ready to fly, the butterflies are removed one by one and examined to verify their sex and if they are perfect. After these proceedings the females are released into the greenhouse, each one with two perfect males.
- As the observed ratio of males and females is at least 5:1 and at most 40:1, there will be males left over depending on the species.
- The remaining males are the physical result of the sustained management plan for the rearing of Lepidoptera and are referred for scientific, commercial and industrial use after death.

Source: RENCTAS – National Network against Trafficking in Wild Animals

MOST FREQUENTLY ASKED QUESTIONS

Jussara Pereira Prado
João Angelo Cerignoni

GLOSSARY

Feromone. Chemical substance attracting animals of the same species.

Osmeterium. Fleshy, tubular, eversible process producing a penetrating odor projected through a slit in the protroraxic segment of certain Papilionidae caterpillars.

Sombrite. Dark plastic screen for sun protection and shading with options of sun filtering.

What is metamorphosis in insects?

The immature stage of an insect species is called larva or nymph, according to the nature of post-embryonic development. Some insects are born with the general aspect of the adult (nymph) and the metamorphosis until the adult stage is normally marked by the development of the wings. This type of metamorphosis is called gradual, simple or incomplete metamorphosis. Holometaboly is the development of insects presenting complete metamorphosis, with four well defined phases: egg, larva (which is an active stage) followed by an inactive stage called pupa from which emerges the adult; such insects are called holometabolous ones. Adult stage may be longer than immature stage, or vice versa, depending on the insect species. For instance, in Ephemeroptera the adult stage lasts one or some days, while the immature, which is aquatic, lasts one year. On the other hand, there is a cicada species whose immature stage lasts seventeen years, although this insect presents incomplete metamorphosis. Generally species with adult stage longer than the immature stage are holometabolous, that is, they present complete metamorphosis.

Notice – the nouns "caterpillar" and "chrysalid" are synomyms of larva and pupa, respectively, when referring to the Order Lepidoptera.

What are Lepidoptera insects?

Lepidoptera insects are represented by butterflies and moths which have colored scales covering the wings. Like all insects, they have the body divided into three parts: head, thorax, and abdomen. The head appendages are 2 antennas, 2 compound eyes, and 0 to three simple eyes and mouth parts modified in a spirotromba which is used to suck liquid food like flower nectaries or juices of fermented fruits. The thorax appendages are the 4 wings and the 6 legs. The reproductive and the excretory systems are located in the abdomen. Butterflies and moths present four stages in their development: egg, larva (caterpillar), pupa (chrysalis) and adult.

What is the difference between butterflies and moths?

Butterflies fly during the day and present the wings covered by bright and varied colors, and clubbed antennae. The moths fly at night and generally have dark colors and threadlike or pectinated antennas. Some species of moths present stinging or venomous caterpillars, but such ones do not occur in butterflies. The pupae of some moth species are protected by a cocoon made by the caterpillar, however butterfly pupae never have this protection and the pupa is called naked one.

How the butterflies reproduce?

Right after leaving the pupa the butterfly is ready to reproduce. The male finds the female being attracted by the pheromones (animal hormones to attract the opposite sex). When the female accepts the male they unite the tips of their abdomen for a while. During this period the male transfer to the female a package of sperma, called spermathophore, fertilizing the eggs.

Why do caterpillars turn into butterflies?

Caterpillar is the life stage in which the insect grows and accumulates food reserves for the adult to reproduce and disperse. Therefore the adult stage is the one in which occurs the production of new individuals and the colonization of new environments.

How long does a butterfly live?

According to the butterfly species this period is variable from two weeks on average, although a Costa Rican species lives two days.

Some tropical species live from six to twelve months. The monarch butterfly which migrates from Canada to Mexico, may live up to six months.

Why do the butterflies have colored wings?

The bright colors may advise the predators, normally birds, that determined butterfly is toxic or unpalatable. Certain butterflies, although not toxic or unpalatable may present colors that imitates toxic species and then escape from the attack of predators. Some butterflies and moths may have colors similar to those of the environment, becoming camouflaged and thus escaping from the predators.

How many butterflies and moths exist?

The scientists believe that there may be circa 12 to 15 thousands of butterflies species and 150 to 250 thousands of moths in the world. Regarding Brazil, more than 3,500 butterflies have already been described. In the Brazilian savanna there occurs circa of one thousand of butterfly species and 5 to 8 thousands of moth species. It is estimated that thousands of species are not described yet.

What is the difference between male and female butterflies?

Many butterfly species present the colors of the male wings different from those of the female wings. However, the more accurate way to separate males from females is to observe the abdomen. Females have rounded abdomen with a round opening in the tip, while the males have a slender abdomen with a small opening in the tip. It is also possible to sex the pupae before adult emergence, but this method is a little more complicated and should be done in laboratory.

Are there butterflies and moths threatened with extinction?

Butterfly and moth caterpillars need specific plants and environments to survive and therefore are especially vulnerable to degradation of the environment. Many Lepidoptera species are in risk of extinction around the world. In Brazil, according to IBAMA, (Ordinances nr.1,522/1989 and nr.45-n/1992), the official list of endangered species of the Brazilian fauna includes several butterfly species, some of which are already considered extinct.

Is it safe to touch a butterfly or a moth caterpillar?

Caterpillars of some moth species have the body covered by stinging hairs that can cause from mild discomfort to minor burns. Most caterpillars are harmless but some such as those of the genus *Lonomia* may cause serious problems that can even lead to death. Pupae and butterflies are extremely delicate and can be easily damaged, but the caterpillars are harmless to the touch. The wings of butterflies and moths are covered by scales. These scales loosen and form a powder that can irritate the eyes, just like any type of dust. The belief that the lepidopteran wings dust may blind is not true.

STINGING CATERPILLARS

Roberto Henrique Pinto Moraes
Instituto Butantan

When dealing with Lepidoptera one should be aware of the sting caterpillars of the families Arctiidae, Limacodidae, Megalopygidae (Figure 9) and Saturniidae (Figures 10 and 11), popularly known in Brazil as "taturanas". *Lonomia obliqua* (Saturniidae)(Figure 10) presents sting caterpillars of gregarious habits. Their bristles are moss green coloring confusing with tree trunks and people may accidentally touch them. After introduction of the bristles on the human skin, a toxin is released that prevents blood clotting and causes hemorrhagic disorders. In more severe cases the venom can lead to death. The symptoms are different from those caused by the "taturanas" of most species. From 1989 onwards hundreds of hemorrhagic cases and deaths due to the contact with *Lonomia obliqua* have been reported in the Brazilian states of Rio Grande do Sul and Santa Catarina where they are popularly known as "taturana" caterpillar, "oruga", or "ruga". In that same period the presence ot *Lonomia obliqua* was also detected in the state of São Paulo.

Main characteristics of *Lonomia obliqua*

Lonomia obliqua caterpillars hatch after 18 days of embryonic development. The first food is the shell of the eggs; then they begin to eat the leaves of the host plants (which can be plants of woods such as cedar, "aroeira", etc, and plants of edible fruits such as avocado, plum, etc). They feed intensely at night and during the day they rest on the trunks. The larval period is 85 days and in it the caterpillars molt their skin (ecdisis) six times. The end of the larval period is characterized by difficulty in locomotion, body bending, loss of bristles and early growth of the cremaster (caudal pointed structure). The insect is now in the prepupa stage. In the pupa phase it does not feed, remaining immobile for 30 days at the base of the trunk of

Automeris sp. (Lepidoptera, Saturniidae) "little pine caterpillar"

Lonomia obliqua (Lepidoptera, Saturniidae) "assassin caterpillar"

Podalia sp. (Lepidoptera, Megalopygidae) "little dog caterpillar".

the trees, protected by plant residues, turning into a male or female moth. The winged ones, after emerging, stretch their wings and are ready for the copulation that will occur in a few days. The fertilized females lay hundreds of eggs, restarting the biological cycle. The moths are ephemeral, for they live 15 days on average and do not feed due to the atrophied oral apparatus. The accidents with Lonomias occur with caterpillars of 4th and 5th instars. At these stages, caterpillars remain during the day in tree trunks, close to the ground and the chance of contact with people is greater. Antilonomic serum, the only drug capable of reversing the haemorrhagic effects of the venom of lonomias, is produced in the Institut Butantan, São Paulo, Brazil, from the bristles of 5th and 6th instars caterpillars. The recognition of caterpillars in these instars is possible due to the following characteristics: gregarious habit, length (approximately 7 centimeters), brown integument, green moss bristles and dorsal spots in semicircles, highlighting a larger white one, close to the head, similar to a letter V or U.

Geographical distribution of *Lonomia* sp. and other poisonous caterpillars

From the genus *Lonomia* eleven species are known, with neotropical distribution. The complete list of American species can be found in Lemaire (2002).

In Brazil *Lonomia obliqua* occurs in the south and southeast, while *Lonomia achelous* inhabits the northern region. In 1994 the presence of *Lonomia achelous* was also registered in Minas Gerais. In the same year, there were hemorrhagic accidents due to contact with caterpillars in the Brazilian states of Goiás and Maranhão, but the species of *Lonomia* could not be identified.

The genera *Automeris* (Figure 11), *Dirphia* and *Hylesia* include several poisonous species and are also present in the Neotropical region. Although megalopigids, arctiids and limacodids occur worldwide, they predominate in the tropical region. In Brazil, they are found all over the country.

Reference

LEMAIRE, C., 2002. The Saturniidae of America. Hemileucinae. Parts A, B, C. Goecke & Evers, Keltern. 1388 p.

Acknowledgements

To those who collaborated for this publication:

Stela Silveira Camargo

Carlos Giovani Tonet

Heraldo Negri de Oliveira

Cynira Any Jovilhana da Silva Gabriel

André Victor Lucci Freitas

Lucas Augusto Kaminski

Jussara Pereira Prado

Roberto Henrique Pinto Moraes